AHA! ACADEMY

MAKING WAVES!

The Physics of **Sound**

Written by Lisa Amstutz

WORLD BOOK

www.worldbook.com

Co-published by agreement between Shi Tu Hui and World Book, Inc.

Shi Tu Hui
Room 1807, Block 1,
#3 West Dawang Road
Chaoyang District, Beijing 100025
P.R. China

World Book, Inc.
180 North LaSalle Street
Suite 900
Chicago, Illinois 60601
USA

Library of Congress Control Number: 2024947135

Aha! Academy: Physics
ISBN: 978-0-7166-7144-2 (set, hard cover)

Making Waves! The Physics of Sound
ISBN: 978-0-7166-7152-7 (hard cover)
ISBN: 978-0-7166-7172-5 (e-book)
ISBN: 978-0-7166-7162-6 (soft cover)

Printed in India by Replika Press PVT LTD, Haryana, India
1st printing January 2025

Staff

Editorial

Vice President
Tom Evans

Editorial Project Coordinator
Kaile Kilner

Senior Curriculum Designer
Caroline Davidson

Proofreader
Nathalie Strassheim

Graphics and Design

Senior Visual
Communications Designer
Melanie Bender

Designer
Shannon Hagman

Digital Asset Specialist
Rosalia Bledsoe

Written by Lisa Amstutz
Advised by Saroj Thapa

Developed with World Book by
Red Line Editorial

Acknowledgments

The publishers gratefully acknowledge the following sources for photography. All illustrations were prepared by WORLD BOOK unless otherwise noted.

Cover: Maksim Kabakou, Adobe Stock; Africa Studio/Shutterstock; Tomas Kotouc, Shutterstock; SeventyFour/Shutterstock; Rudmer Zwerver, Shutterstock

Julian Herzog (licensed under CC BY 4.0) 25; SDI Productions/iStock 41; Library and Archives Canada 10; Library of Congress 31; NASA 7, 47; NASA/SOFIA/Lim, De Buizer, & Radomski et al.; ESA/Herschel; NASA/JPL-Caltech 21; Omegatron (licensed under CC BY 2.0) 15; Yan Krukau, Pexels 4; Public Domain 23; Public Domain (American Institute of Physics) 20; Public Domain (Isaac Newton Institute) 19; Science Museum Group (licensed under CC BY 4.0) 42; Spyrogumas (licensed under CC BY 3.0) 15, 48; Shutterstock 3, 4, 5, 6, 7, 8, 9, 10, 11, 12, 13, 14, 15, 16, 17, 18, 19, 20, 21, 22, 23, 24, 25, 26, 27, 28, 29, 30, 31, 32, 33, 34, 35, 36, 37, 38, 39, 40, 41, 42, 43, 44, 45, 46, 47, 48; U.S. Air Force 23

There is a glossary of terms on page 48. Terms defined in the glossary are in type that looks like *this* on their first appearance on any spread (two facing pages).

Contents

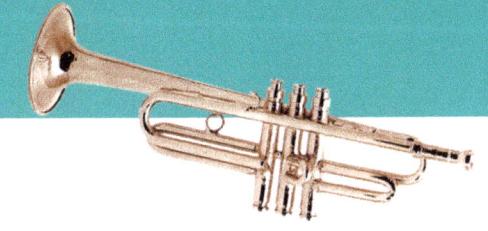

Introduction

The school bell rings. *Brrring!* Lockers slam. *Bang!* Outside the window, birds chirp. Sounds are all around us! But did you ever stop to think about how they work? Why is a sound's **pitch** high or low? Why are some sounds loud and others quiet?

Sound is a type of energy that moves in waves (not the kind you surf!). Sound waves are molecules moving in a **medium,** whether air, water, or even stone. Our ears pick up these vibrations and translate them into sounds. These sounds help us interpret the world around us. They warn us of danger. They help us to communicate.

Why do birds sound different from the noises at school? Physics will help us find out.

MOLECULES IN MOTION

Sound can travel through any *medium* that contains matter. Air, water, glass, wood, metal, and stone can all carry sound. Some mediums conduct sound waves better than others. The only place sound cannot travel is in a *vacuum*—like space. A vacuum has little or no matter in it, not even air! Sound cannot travel in a vacuum because there are no molecules to move.

Move it, molecules!

Sound starts with a force that sets the molecules in motion. When you snap your fingers or bounce a basketball, that movement pushes molecules out of the way. Those push other molecules in front of them, creating a wave.

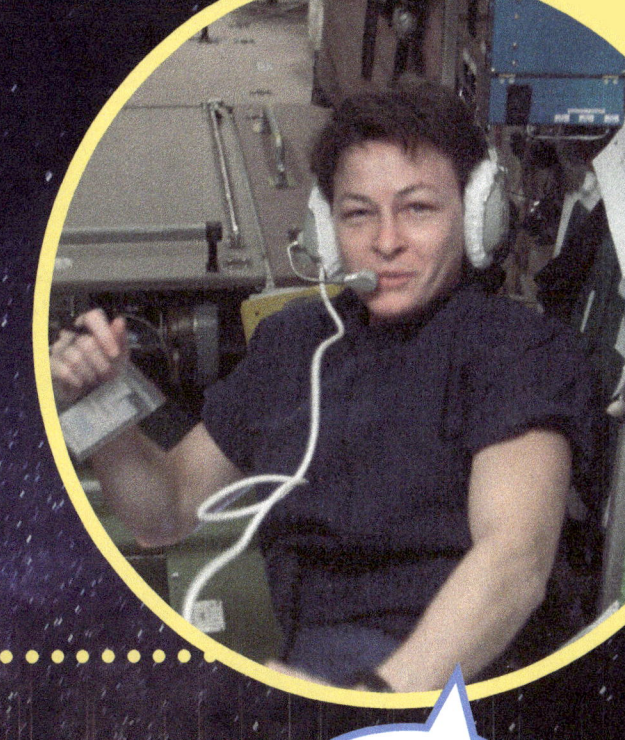

The 1979 horror movie *Alien* had the tagline "In space, no one can hear you scream." But is this true? The answer is yes— at least, not without the help of technology. There are simply too few molecules in space to carry sound. But astronauts can talk to one another using radio waves. They can also use hand signals if their technology fails.

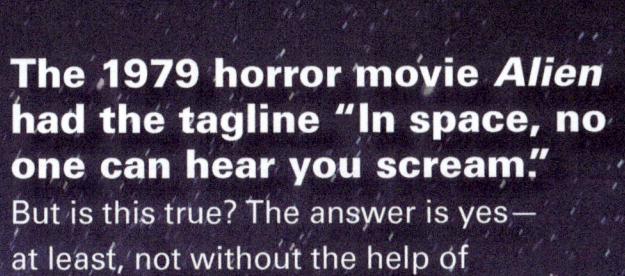

Can you hear me now?

DID YOU KNOW?

You probably have vacuums around your home! The inside of an incandescent light bulb is a vacuum. So is the area between the layers of insulation in a thermos or water bottle. And a vacuum cleaner uses suction to create a vacuum.

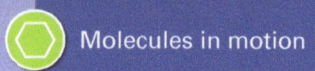
Making waves

It takes a lot of energy to create all that sound. The energy from the drummer's arms brings the drumstick down. When it hits the drum, the energy transfers through the stick and into the drum. The drumhead begins to vibrate, creating sound energy. Those vibrations push the air molecules around them.

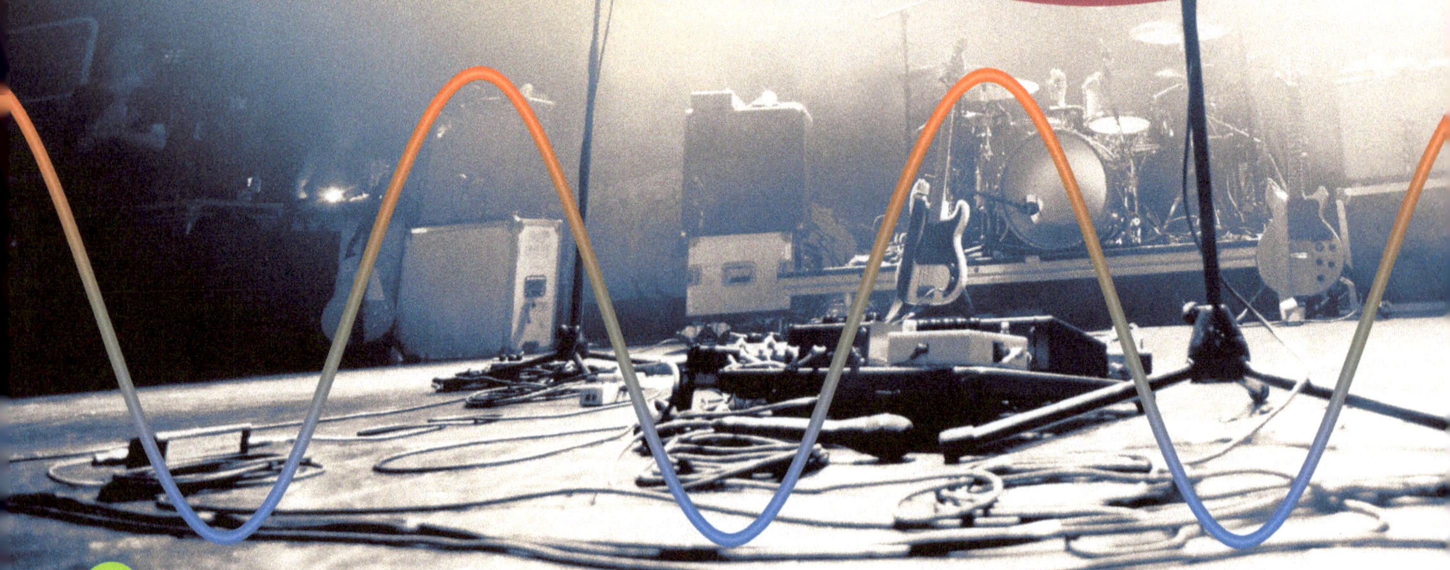

Compression

You're at a concert, watching your favorite drummer. Her arms move so fast they're almost a blur! Each drumroll fills the air with sound.

> Here, drum, have some sound energy.

When the drumhead moves up, the air molecules around it bounce off faster than usual. They are pushed closer together; this is called *compression*. When the drumhead moves back down, the air molecules bounce off with less energy. This is called *rarefaction*. A wave cycle consists of one compression and one rarefaction. On a graph, these look like mountains and valleys.

Rarefaction

Another way to think about a wave cycle is to picture a Slinky—that metal spring toy that can walk down stairs. If you give one end a push, a wave travels through the metal spring. The coils get closer together at the front of the wave and farther apart behind it. Sound waves work in the same way.

Turn up the volume

Dynamite explodes with lots of energy! It causes waves with huge highs and lows. A whisper, on the other hand, contains very little energy. It makes small waves that stay close to the center line.

amplitude

I invented the telephone.

Decibels are named for **Alexander Graham Bell**. Bell invented both the telephone and the audiometer, a tool used to measure hearing loss. Bell was born in Scotland in 1847 and died in Canada in 1922. His mother was nearly deaf, and Bell worked as a teacher of the deaf, later marrying one of his former students. So it is no wonder he was fascinated with sound!

A jackhammer outside your window can be earsplitting, while a bird sings quietly. What makes one sound loud and the other soft? Two things: the *amplitude* and the *intensity* of the waves. On a graph, amplitude is the height of a wave. It measures the energy of the wave. Picture an ocean wave crashing on a beach. The taller the wave, the more energy it contains. The same is true for sound waves.

Intensity also affects the loudness of a sound. Intensity is the amount of energy transmitted through an area in a certain time in the direction of the wave. A sound gets less intense as you move away from it. That is why a siren can be deafening up close but very faint if you are far away.

There is such a thing as too much quiet! A specially built chamber at Orfield Laboratories in Minnesota, United States, is the quietest place on Earth, according to the Guinness Book of World Records. It is used to test products and even train astronauts. Without any *echoes*, people find it hard to balance.

We measure amplitude in decibels (dB). This chart shows where common sounds fall on the decibel meter. Sounds above 80 dB can damage your hearing. Those above 120 dB can cause physical pain. Ouch!

Safe							Unsafe						
10	20	30	40	50	60	70	80	90	100	110	120	130	140
Breathing	Ticking watch	Whisper	Quiet library	Rain	Voices talking	Vacuum cleaner	Police siren	Hair dryer, lawn mower	Dog's bark	Sports event	Concert	Jackhammer	Fireworks

Pitch **perfect**

A sound's pitch is caused by its _frequency_. This number tells us how often the sound's waves pass in a given amount of time. If one wave passes per second, the frequency is 1 hertz (Hz). If two waves happen per second, the frequency is 2 Hz, and so on. The time (in this case 1 second) is known as the _period_ (T).

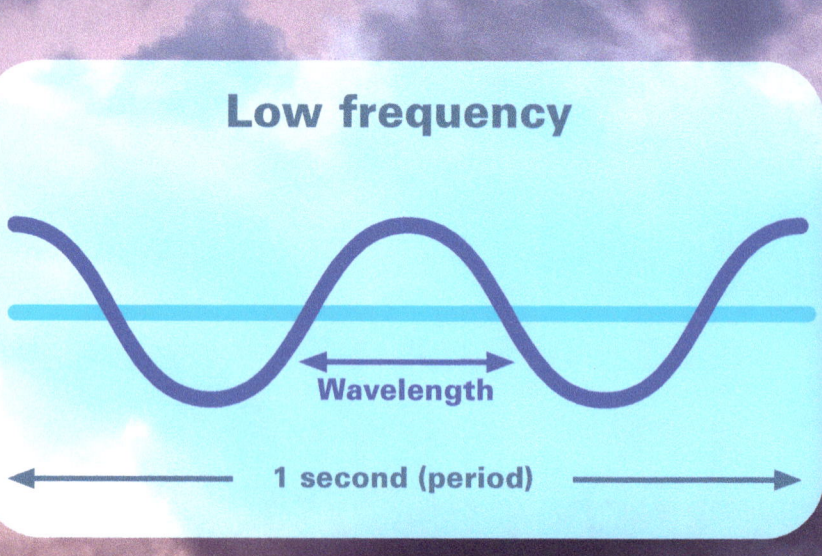

Low frequency

Wavelength

1 second (period)

Tra-la-la!

Whales talk to one another using long, low sounds. These can travel long distances though the water. Whales vary the frequency of these sounds, so the pitch changes. These different pitches form a song. Whales may repeat a song over and over.

It's a dark and stormy night. You hear a low rumble of thunder. Then you hear a sharp, high-pitched scream. Aaaaaah! Both sound scary! But their *pitch* is very different. Pitch is another way to describe sound.

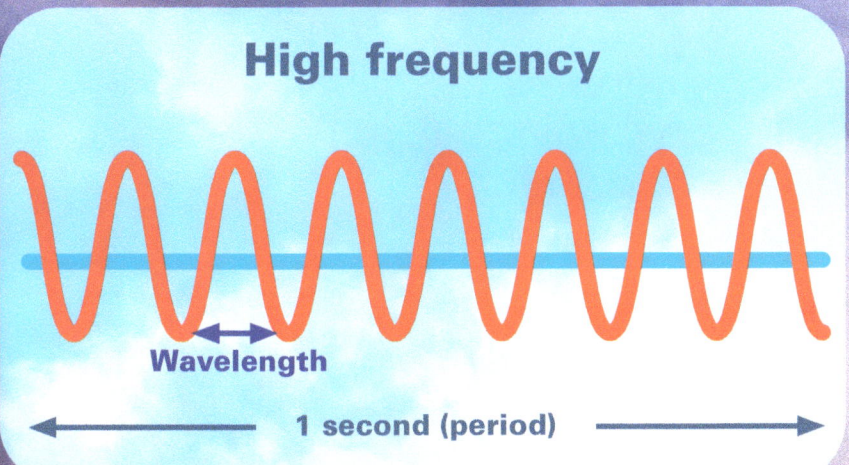

High frequency

Wavelength

1 second (period)

Wave cycles that are close together have a high frequency because the source of the sound is vibrating fast. They make a high-pitched sound. Wave cycles that are farther apart have a lower frequency. They make lower sounds.

CURIOUS CONNECTIONS

A violin string is short and thin. It is stretched tightly over the violin. Waves move quickly through it. When you pluck the string, it makes a high-pitched sound. A string bass has longer, thicker strings. They make low-pitched sounds.

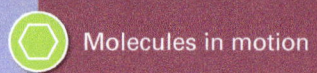

Looking at **sound**

An oscilloscope is a tool used to measure electrical signals.
It can show sound waves too. A microphone turns a sound into electrical signals.
When the microphone is plugged into the oscilloscope, these signals show up as
waves on the screen. The image is called an oscillogram.

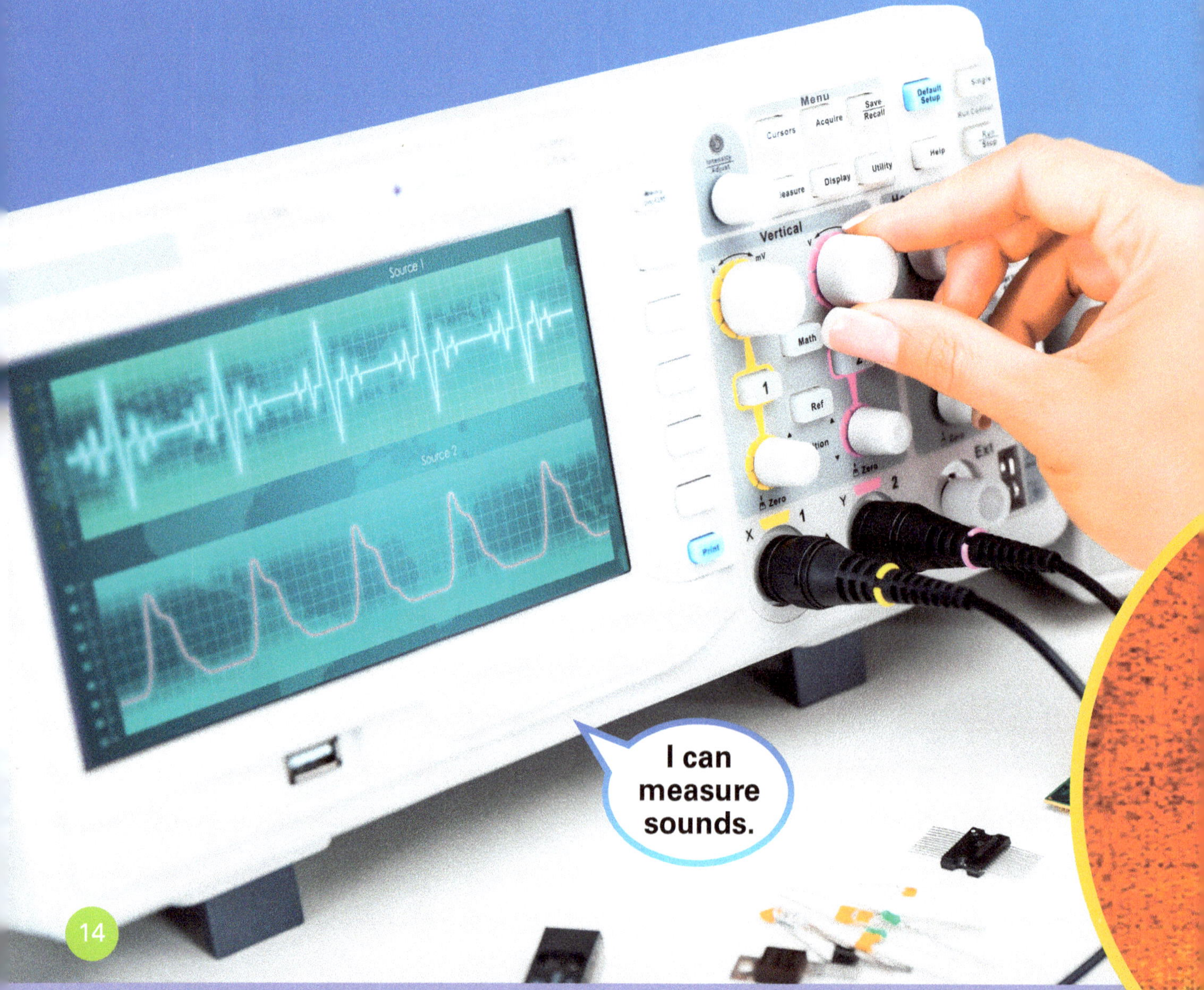

I can
measure
sounds.

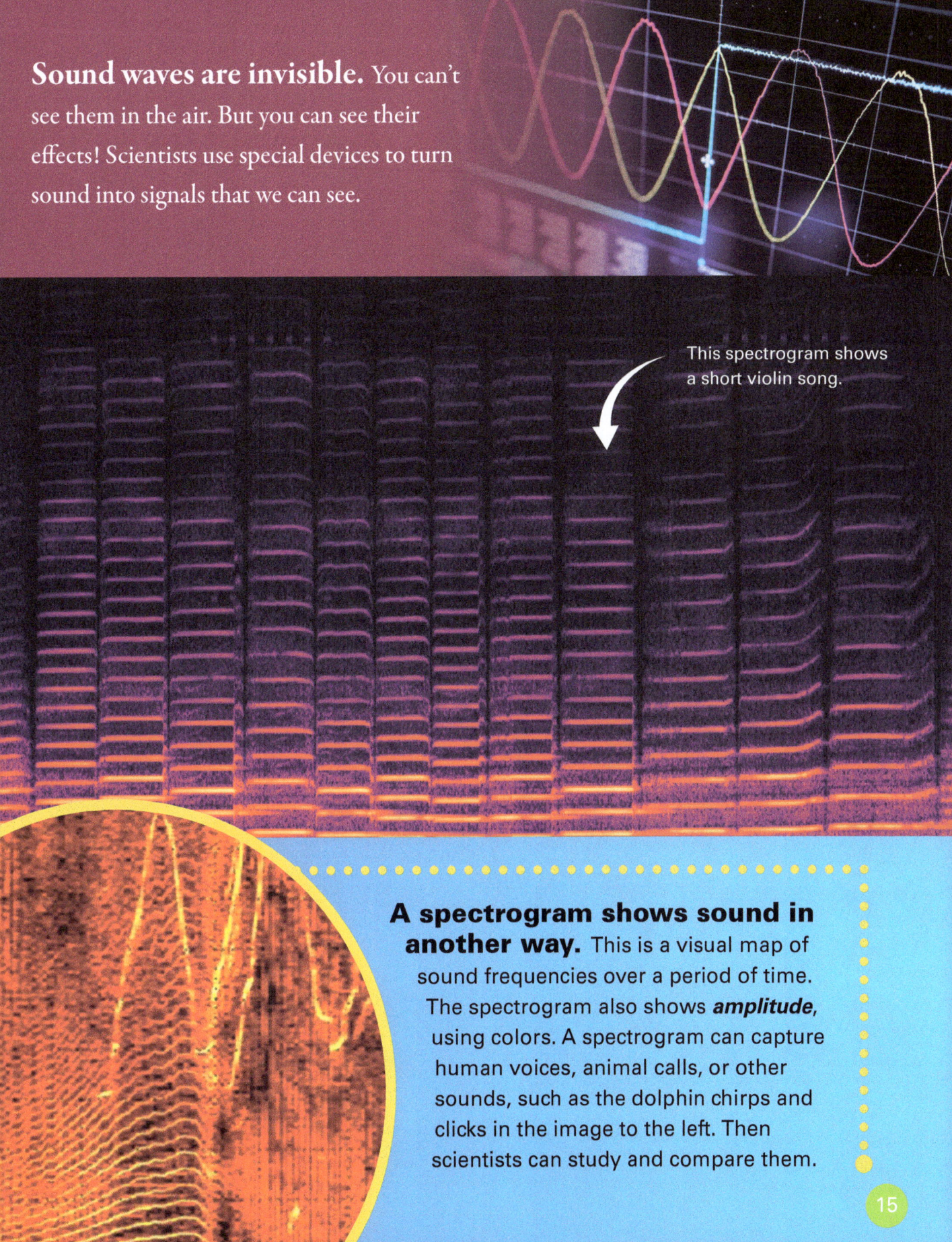

Sound waves are invisible. You can't see them in the air. But you can see their effects! Scientists use special devices to turn sound into signals that we can see.

This spectrogram shows a short violin song.

A spectrogram shows sound in another way. This is a visual map of sound frequencies over a period of time. The spectrogram also shows *amplitude*, using colors. A spectrogram can capture human voices, animal calls, or other sounds, such as the dolphin chirps and clicks in the image to the left. Then scientists can study and compare them.

HOW WAVES BEHAVE

Sound waves can behave in surprising ways. They can bounce and even bend. This changes the way they sound to the listener.

Helllllooooooooo! Have you ever heard your voice *echo* in a big room? When waves hit a wall or a cliff, they bounce back. The reflected sound is called an echo. The waves lose some energy when they hit the wall. That makes the echo softer than the original sound.

○ **Bat sonar waves**

◌ **Reflected sound waves**

TECH TIME

One way humans use echolocation is with a technology called sonar. Sonar stands for *s*ound *n*avigation *a*nd *r*anging. Scientists use sonar to explore the depths of the ocean, where light cannot reach. Ships and submarines can use sonar to navigate.

If a sound and its echo are less than one-fifteenth of a second apart, our ears cannot hear the difference. So you will not notice an echo unless you are far enough away to hear both sounds.

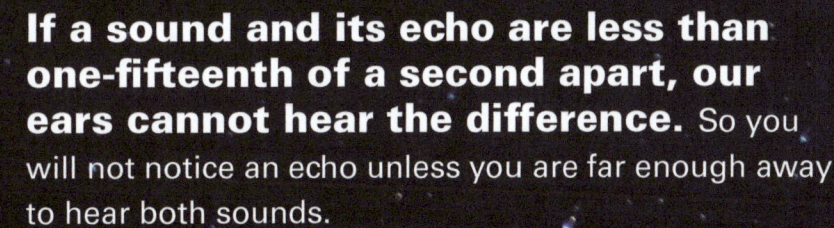

Watch out for that tree!

Bats hunt in the dark, so they can't rely on their eyes to spot prey. Instead, they send out high-pitched clicks. Then they listen for the echo. It guides the bat to its food. This is called *echolocation*. Some whales use echolocation too. It can even help humans! People who are blind can sense objects around them by making a sound and listening for the echo.

Sound waves generally travel in a straight line.
But sometimes they can bend around a corner or through a doorway. Then they spread out. That process is called *diffraction*.

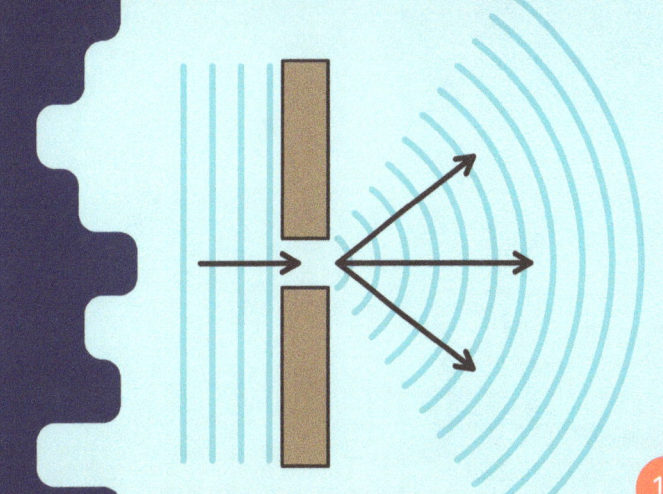

Sound on the **move**

The speed of sound depends on several factors. The first of these is the *medium*. You might be surprised to learn that sound travels faster in water than in air. And it moves even faster in a solid, such as wood. In general, sound moves fastest through solids, next fastest through liquids, and slowest through gases. This is because the molecules in solids are packed close together. They are looser in liquids, and spread out widely in gases.

Speed of sound

	Speed in feet per second	Speed in meters per second
Steel door	17,100	5,200
Aluminum foil	16,000	5,000
Glass window	14,900	4,540
Maple tree	13,480	4,110
Brick wall	11,980	3,650
Warm seawater (77 °F/25 °C)	5,023	1,531
Air (59 °F/15 °C)	1,115	340

Have you ever tried to listen to someone talk underwater? Or through a wooden door? Which sound do you think would reach you faster?

Temperature also affects the speed of sound. Warmer molecules have more energy, so they can move around faster. That means sound travels faster on a hot day. It travels faster in tropical oceans than in icy seas.

Sir Isaac Newton used an *echo* to measure the speed of sound. He stood in a hallway and stamped his foot. He timed how long it took to hear the echo. Because he knew how long the hallway was, he could determine the speed of sound. His answer was very close to what we have measured with more modern instruments!

In the real world, sound waves often run into bodies of water, walls, trees, or other objects. When sound waves move from one medium to another, they bend and change speed. This is called *refraction*. Imagine a car that drifts off the road into gravel or mud. The wheels on that side slow down and pull the car in that direction. Refraction works the same way with sound waves.

Sound moves pretty fast!

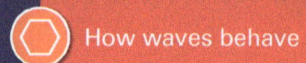

Special effects

Because the source of the sound is moving, the sound waves in front of it get bunched up. The ones behind it get spread out. That makes the *pitch* sound different to someone who is standing still.

**Small wavelength
High frequency**

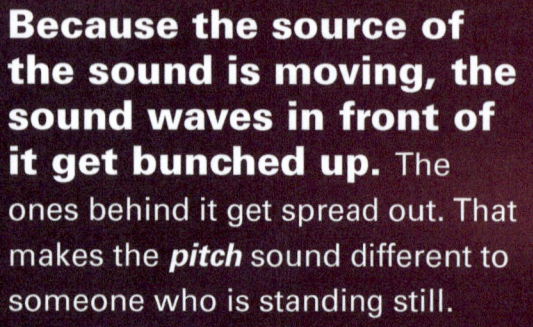

 I knew it!

The Doppler effect is named after physicist Christian Doppler, who first described it in 1842. Doppler taught at the University of Vienna in Austria. He predicted that the Doppler effect would help astronomers track the movements and distances of stars. It took 100 years, but he was right!

Did you ever notice that a siren's pitch sounds higher as it approaches and then lower once it has passed? That is because of the Doppler effect. The *frequency* of the sound does not actually change. But it seems like it does.

**Long wavelength
Low frequency**

AMBULANCE

CURIOUS CONNECTIONS

SPACE The Doppler effect also happens with light waves. The light from a star looks redder if the star and Earth are moving away from each other. It looks more violet if they are moving toward each other. This is known as the Doppler shift.

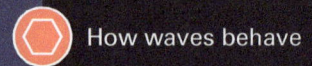

Breaking the sound barrier

Just as a boat creates ripples in the water around it, an aircraft creates ripples in the air. At supersonic speed—faster than the speed of sound—these ripples compress into a cone shape at the back of the plane. This is known as a Mach cone.

Most fighter aircraft can reach supersonic speeds. But they only do so in certain places, such as open water that is at least 15 miles (24.1 km) from shore, and above 10,000 feet (3,048 meters) if possible. That keeps them from breaking glass or scaring people and animals.

Sound travels through the air at about 767 miles (1,234 kilometers) per hour. But some aircraft fly even faster! When an aircraft travels faster than the speed of sound, it compresses the air molecules in front of it. After it passes those air molecules, they expand quickly, causing a sudden change in air pressure. This creates a loud noise called a *sonic boom*. This boom can startle people and even break windows.

Now, that's fast!

Pilot Chuck Yeager was the first person to break the sound barrier. A fighter pilot during World War II, he later became a test pilot. Yeager flew the secret X-1 aircraft. In 1947, it was the first plane to reach supersonic speed. It flew at 662 miles (1,065 km) per hour at 40,000 feet (12,000 meters). In this case, sound traveled slower than its average speed. Altitude and temperature can affect the speed of sound.

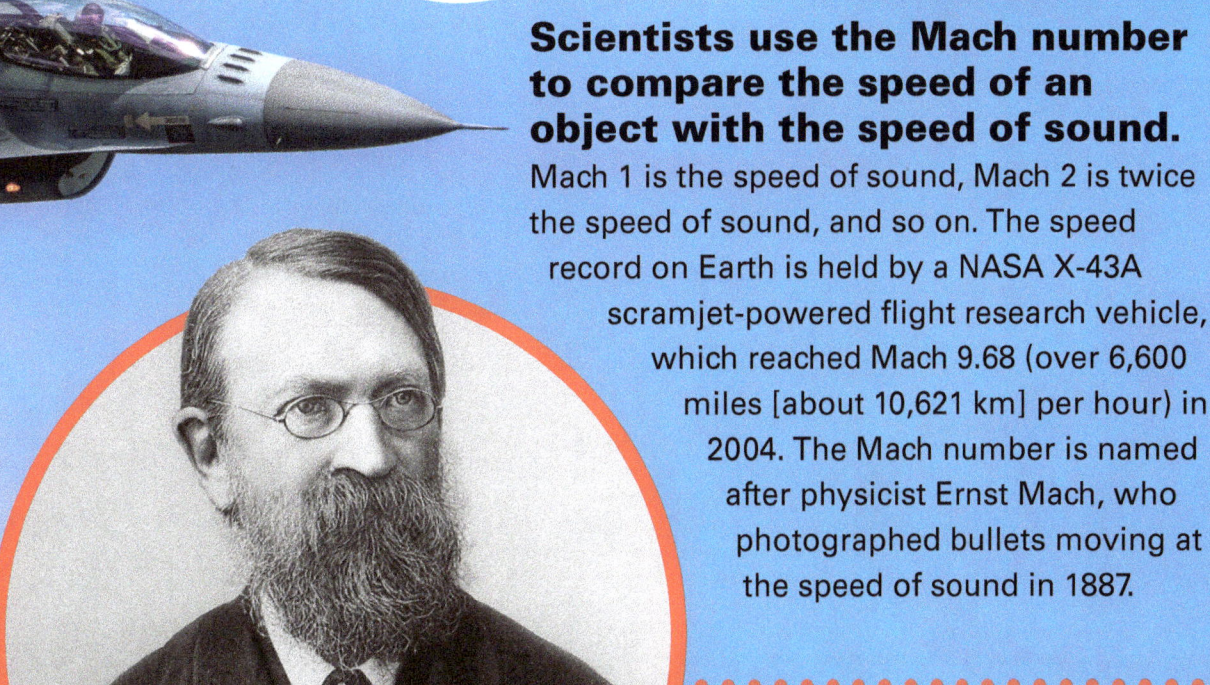

Scientists use the Mach number to compare the speed of an object with the speed of sound. Mach 1 is the speed of sound, Mach 2 is twice the speed of sound, and so on. The speed record on Earth is held by a NASA X-43A scramjet-powered flight research vehicle, which reached Mach 9.68 (over 6,600 miles [about 10,621 km] per hour) in 2004. The Mach number is named after physicist Ernst Mach, who photographed bullets moving at the speed of sound in 1887.

SOUNDS ALL AROUND

3

Before machines, the only sounds were made by human voices or nature. Now, our world is filled with the sounds of technology, from jet engines to TV's to dishwashers. At low levels, they may be annoying. At high levels, they can cause mental and physical harm.

Have you ever noticed how quiet it becomes when the power goes out? Until it's gone, we often don't notice the low-level background noise in our homes. This constant hum includes sounds from refrigerators, air conditioners, fans, and other devices. The typical living room has a noise level of 40 decibels. A clothes dryer can hit 58 decibels, while a *vacuum* cleaner can reach 89 decibels.

TECH TIME

Ever wonder how noise-canceling headphones work? They actually use sound waves to protect your ears! The material they're made of blocks some noise. But the headphones also contain a microphone that picks up noise around it. Electronics measure the *frequency* and create sound waves that are similar—but out of sync. The highs and lows of this "antinoise" cancel out the sound waves.

Electric and hybrid vehicles pose a unique problem: at low speeds, they can be too quiet!

These vehicles emit one-tenth of the noise of cars with gas or diesel engines. This is great for cutting down on sound pollution. However, the silence can be dangerous to people who don't hear them coming. So automakers add a low-level hum to these machines to alert pedestrians.

Shhh!

A noisy world

Keep it down out there!

Noise can even cause physical damage. Sound waves pass through the body like they do through any other *medium*. At very high decibel levels, they can cause joint damage or affect the lungs or eyes. They can also damage hearing. Theoretically, sound could kill. But it would take a noise level of around 240 decibels, which is nearly impossible to generate.

When sounds are annoying or harmful, they are called noise. A noisy airport or highway can make humans stressed. This kind of noise pollution can stress animals too. It may keep them from finding food or mates.

Many people live in areas where there is noise pollution. What can be done? In some areas, sound barriers along highways block sound from reaching nearby homes. Trees and buildings can help too. Indoors, heavy curtains and white-noise machines can muffle noise.

TECH TIME

Wondering how loud a concert or school cafeteria is? There's an app for that! Smartphone apps can measure the loudness of sounds around you. They will tell you if a sound is loud enough to damage your hearing.

The highs and lows

What frequencies can animals hear?

This graph shows the range of frequencies each of these animals can hear.

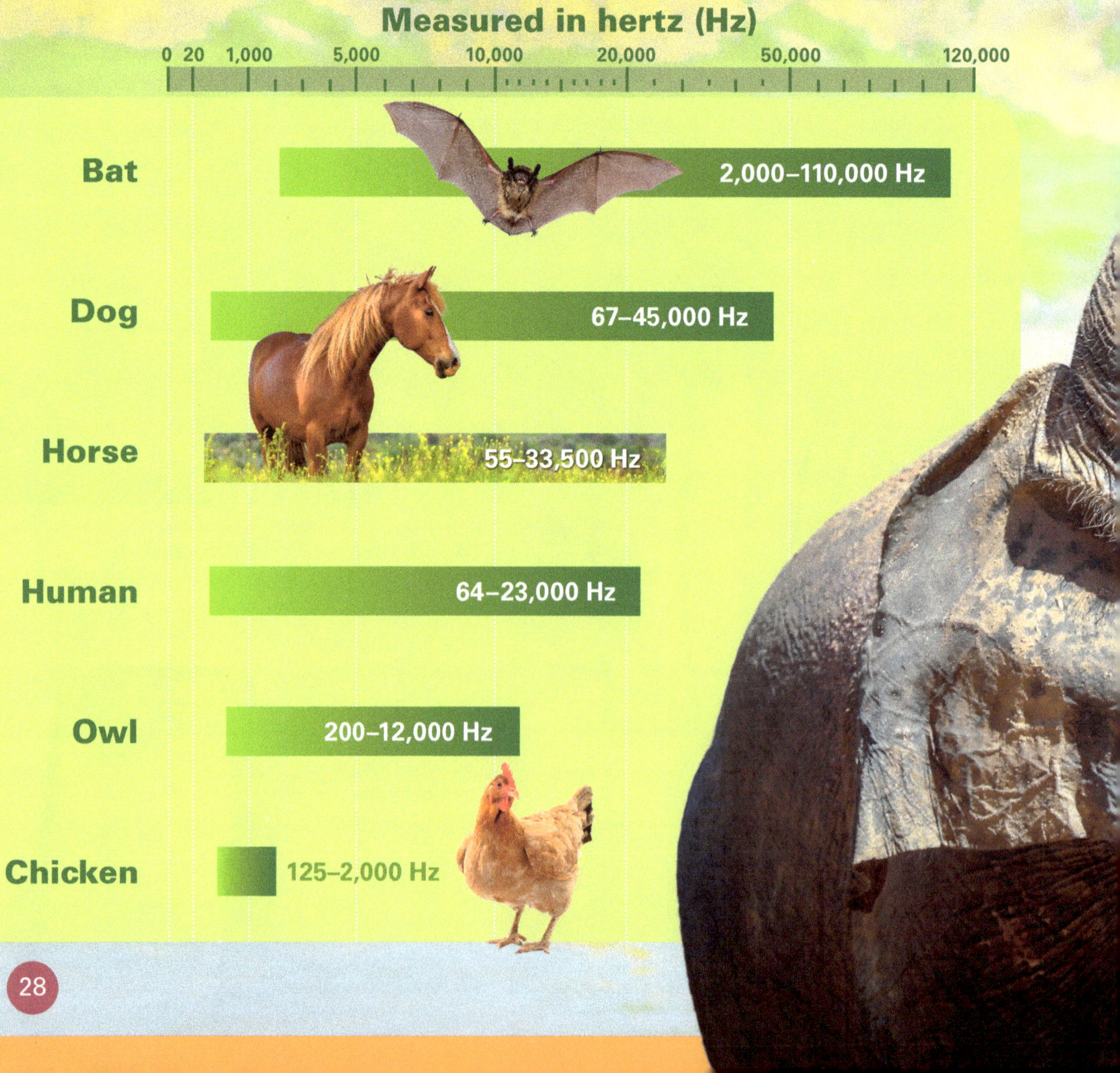

Measured in hertz (Hz)

0 20 1,000   5,000   10,000   20,000   50,000   120,000

Bat 2,000–110,000 Hz

Dog 67–45,000 Hz

Horse 55–33,500 Hz

Human 64–23,000 Hz

Owl 200–12,000 Hz

Chicken 125–2,000 Hz

Some sounds are too high or too low for humans to hear. We can hear sounds only between about 64 Hz and 23,000 Hz. Frequencies higher than that are known as ultrasound. Humans cannot hear them, but some animals can. A dog whistle emits sound that is about 35,000 Hz. That sounds silent to you, but your dog can hear it. Dogs can also hear sounds that are too soft for humans to hear.

Frequencies too low for humans to hear are called infrasound. These range from about 0.1 Hz to 20 Hz. Many large animals make sounds that fall into this range, including elephants, tigers, and alligators. Lightning, earthquakes, volcanoes, and waterfalls can also make sounds in this range. These low sounds can travel long distances.

TECH TIME

Even though we can't hear it, ultrasound is often used in health care. Ultrasound machines send sound waves through a person's body and record the *echoes*. The results are turned into an image. It lets doctors "see" inside the body.

I make sounds you can't even hear.

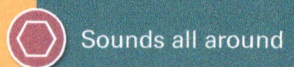
The sound of technology

From TV's to computers to phones, devices constantly bring sound into our homes, workplaces, and schools. But those sounds are produced far away. How can they travel to our homes? The trick is in converting them to digital signals—and back again.

A microphone captures the sound waves at their source. It turns them into an electrical current. An analog-to-digital converter samples the current and records the signals as a series of ones and zeros. These bits of digital data can be stored on a computer, phone, or other device. Or they can be sent through wires or across radio waves. When they reach the device on the other end, they are turned back into vibrations that we hear as sound.

I recorded sounds before you were even born!

In 1877, **Thomas Edison** came up with a brilliant idea—a way to record sounds and then play them back. He called his invention the phonograph. The machine used a thin disc, or diaphragm, attached to a stylus, or needle. Sounds made the diaphragm vibrate, and the stylus made grooves in a foil-wrapped cylinder. The process could then be reversed. The stylus moved over the grooves, creating vibrations in the diaphragm that created sounds.

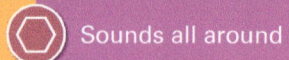
Sound **innovations**

When you go to a play or concert, take a look around the auditorium or theater. These rooms are specially designed to funnel sound to the audience. They keep it from *echoing* or bouncing around. Sound baffles, or panels made of foam or fiberglass, hang from the ceiling to diffuse sound, and wall coverings help absorb sound. Even the shape of the room is important. Auditoriums often slant up toward the back.

Sound technology has exploded in recent decades. We've come a long way since that first phonograph! Technology has improved the quality of sound in many areas, such as live performances, movies, and video games.

One growing field is the use of virtual or augmented reality.

By wearing a headset or other device, a user feels as if they are surrounded by a three-dimensional world. Adding 360-degree sounds to this experience makes it so realistic, you'll think you're actually there! These sounds are recorded using an ambisonics microphone, which can record sound from multiple angles.

Whoa, where am I?

CAREER CORNER

A Foley artist creates sound effects for films and TV. Named for creator Jack Donovan Foley, a Foley artist adds sounds after a scene is filmed to make it feel more realistic. These might include creaking stairs, swishing fabric, or breaking glass. Some of these sounds are recorded from real actions, like the sound of a doorbell ringing. Others are created in the studio. A Foley artist might flap leather gloves to mimic the sound of birds' wings or walk up and down stairs with special shoes to create the sound of footsteps.

4

LET'S TALK

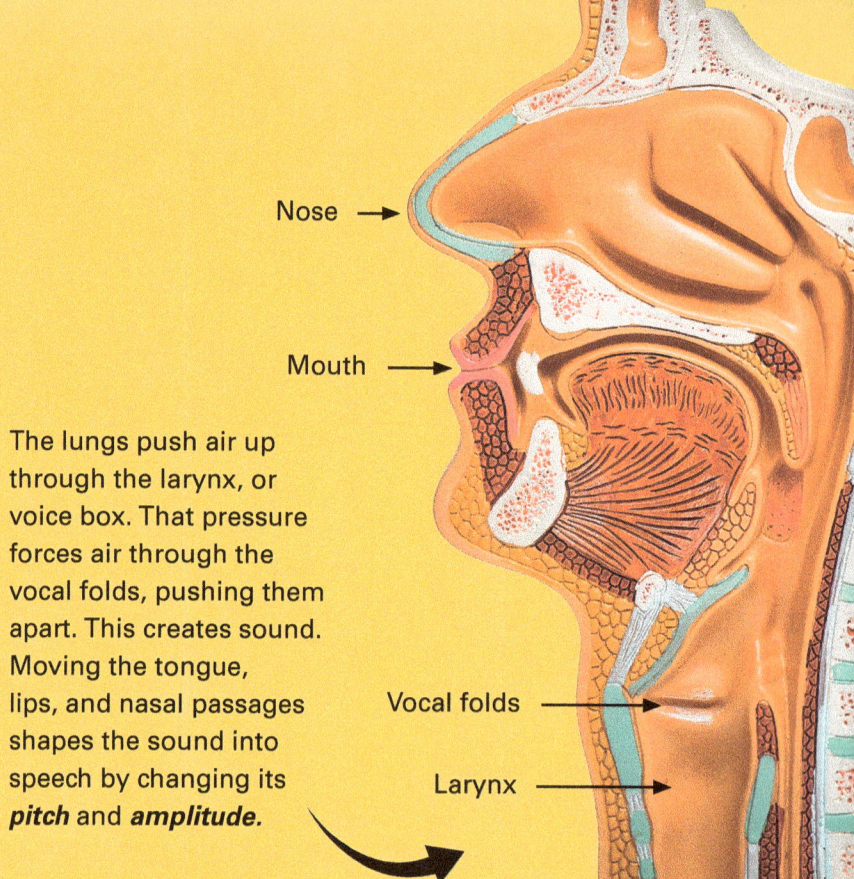

Nose →

Mouth →

The lungs push air up through the larynx, or voice box. That pressure forces air through the vocal folds, pushing them apart. This creates sound. Moving the tongue, lips, and nasal passages shapes the sound into speech by changing its *pitch* and *amplitude.*

Vocal folds →

Larynx →

Your voice resonates in the space in your mouth.

Hola! Bonjour! Nǐ hǎo! Jambo! Hello! Whatever the language, speech is an important way in which humans communicate. But speech is a complex process. Several body parts must work together to create sounds.

Sounds really resonate with me.

A guitar or piano has open spaces inside where sound can resonate. It bounces off the walls of the space, making it sound louder or deeper. Similarly, the human voice resonates inside the body. The primary places this happens are the oral cavity, the pharynx, and the nasal cavity.

CAREER CORNER

Some people have trouble forming words correctly. A speech therapist can help by providing tools and exercises to help people speak and use language skills. For example, the therapist and patient might work on exercises to strengthen the mouth and tongue. They might practice words or play word games together.

Getting an **earful**

The ear is a complex organ. It holds the tiniest bones in the body. The ear has three main parts: the outer ear, middle ear, and inner ear. The outer ear catches sound waves. It funnels them into the ear canal. Then the sound waves hit the eardrum. This thin *membrane* is about the size of a small button. It begins to vibrate.

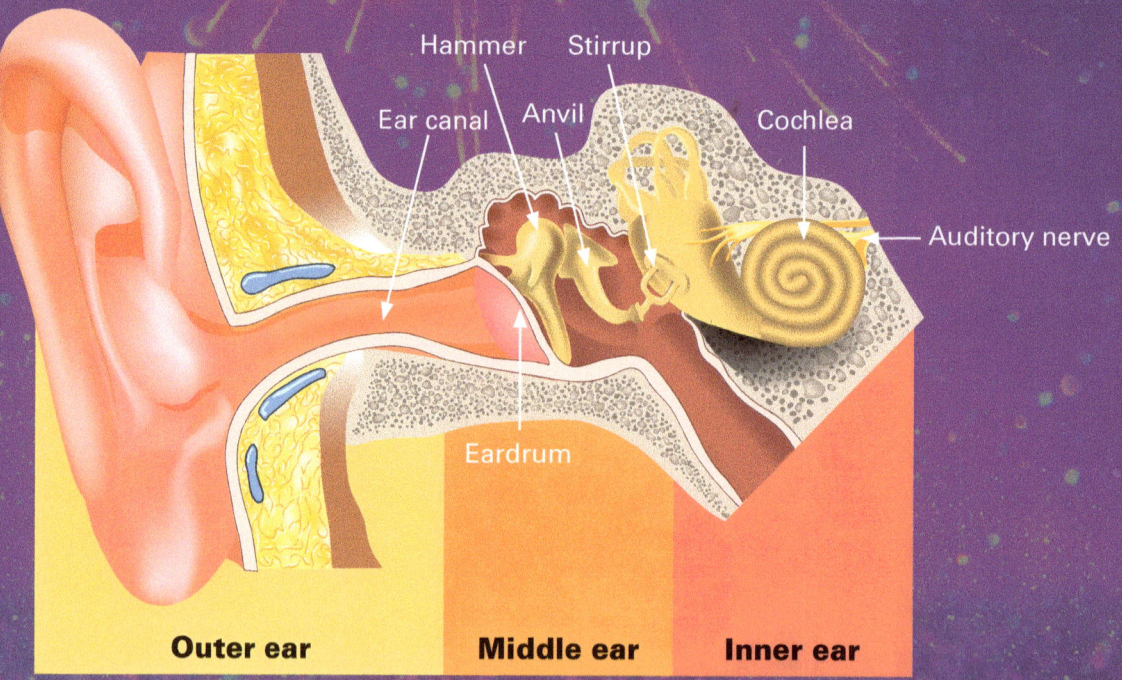

Hammer Stirrup
Ear canal Anvil Cochlea
Auditory nerve
Eardrum

Outer ear **Middle ear** **Inner ear**

The eardrum vibrates against three tiny bones: the hammer, anvil, and stirrup. These carry the sound to the cochlea. This pea-sized organ looks like a snail shell. It is filled with fluid, which begins to ripple. The fluid moves bundles of tiny hair cells. Ions—atoms carrying electrical charges—move through these bundles and release chemicals. These trigger electrical signals that travel through the auditory nerve to the brain.

The human voice can make an amazing range of sounds, from grunts to screams to songs. The human ear is perfectly designed for collecting these sounds and transmitting them to the brain.

Sounds that are too loud or last too long can damage hearing. Turning on your headphones at full blast can cause hearing loss in just a few minutes. Other loud sounds, such as music at concerts, sirens, and fireworks, can damage your hearing quickly too. Be sure to wear earplugs or, if you can, move away from extremely loud noises!

I said, turn it down!

CAREER CORNER

Sound designers create sound effects and even music for games and apps. They record, edit, and synthesize sounds to make games feel realistic. Audio engineers record, mix, and edit sounds. If you like technology, this might be the career for you. Audio engineers work in theater, film, sports broadcasting, podcasting, and many other fields.

Voices around the world

Speech sounds vary around the world too. In some African languages, clicks are used as consonants. These languages are known as "click languages." Other cultures use whistled languages. Shepherds in the Canary Islands use a language called Silbo to talk to each other across mountain valleys. Whistled languages are most common in mountainous areas or dense forest. The sound can carry up to 10 times farther than a shout.

Some cultures use their voices to make unique sounds. From music to daily communication, keep an ear out for these meaningful tones!

In parts of Asia, Canada, and South Africa, some people throat sing. Using this technique, a singer produces more than one *pitch*, or note, at the same time! They do this by changing the shape of their larynx, lips, and tongue.

 # CURIOUS CONNECTIONS

MUSICAL THEATER

Can an opera singer really shatter a glass with her voice? In theory, yes! Resonance happens when the *frequency* of a vibration matches the natural frequency of an object. This is the frequency at which the object tends to vibrate when it is hit, plucked, or strummed. When the frequency of a singer's voice matches the natural frequency of a glass, the glass vibrates harder. If she sings that note loudly enough for a long time, the intense vibration can weaken the glass and make it break.

The sound of music

String instruments use strings vibrating across a hollow wooden body to create sound. The musician may pluck, strum, or run a bow across the strings. Back in the 500's B.C., Greek mathematician Pythagoras noticed that the length of a string affected its tone. This became known as the first law of strings.

> I'm pickin' up good vibrations.

Most woodwind instruments, such as clarinets and oboes, make sound by vibrating columns of air. The player blows air across a thin piece called a reed and covers holes in the instrument to change its pitch. Such brass instruments as trombones or trumpets are similar—players vibrate their lips across the opening to create sound that resonates through the body of the instrument.

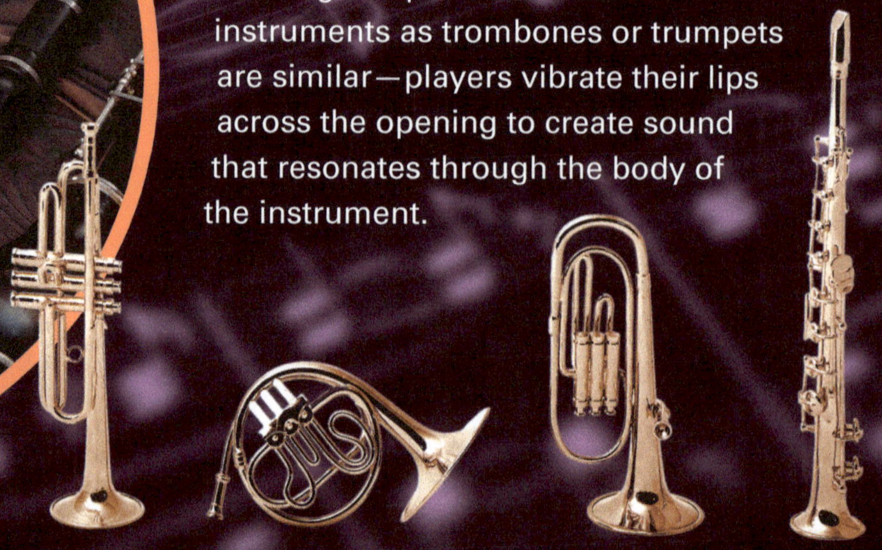

Music is made by combining sounds of different *pitches,* lengths, and *amplitudes* in a pleasing way. People make music by singing or by playing instruments.

Percussion instruments are hit with a stick, hand, or other object to create sound. Most provide a steady rhythm to music. Some, like bells or a xylophone, can play notes too.

CAREER CORNER

Music sounds pleasing to the ear. But it can also heal! Music therapists use music to help relieve stress, improve mood, aid in communication, and even improve health for patients with dementia, stroke, and other diseases. Patients may listen to or create music during therapy sessions. Many colleges offer degrees in music therapy.

A piano fits into two categories. Because it has strings inside, it is a string instrument. But those strings are hit by hammers. That makes it a percussion instrument too!

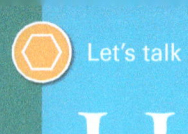

Hearing the future

> I did not fit in anyone's ear.

For severe hearing loss, a cochlear implant may be used. Part of the device is implanted under the skin behind the ear. Tiny wires carry the sound into the inner ear. Another part is worn on the outside. These devices do not amplify sound. Instead, they bypass the damaged parts of the ear and send signals directly to the auditory nerve. Wearers learn to identify these signals as sound.

Long ago, people used ear trumpets to help them hear better. These large, funnel-shaped devices helped direct sound into the ear. Today's technology is far more powerful. Hearing aids are tiny devices that fit inside the wearer's ear. They use digital technology to amplify sounds and can even connect to other devices via Bluetooth. They suppress and control background noise as well.

If the middle or inner part of one ear is blocked or too damaged, doctors can use a bone conduction implant. This implant carries sound through the patient's bone to the good ear. This technology is also used in certain types of headphones.

DID YOU KNOW?

Modern hearing aids can connect to a person's phone or television through Wi-Fi. That means the connected person can be the only one hearing the conversation or TV show.

Sound in action!

You will need:

- Glass bowl
- Piece of plastic wrap
- Large rubber band
- Handful of candy sprinkles
- Four to six items you could use to make sounds
- Music player with speaker

Give it a try

1. Lay plastic wrap over the top of the bowl.
2. Use the rubber band to hold the plastic wrap in place.
3. Place a handful of candy sprinkles on the plastic wrap.
4. Now try making sounds near the bowl. Record your observations in the chart to the right. You might try humming near the bowl at a higher or lower pitch. Perhaps you could set a speaker close to the bowl and play music. Or tap a metal pan with a spoon. What other sounds could you try?

Try this next!

What would happen if you put a speaker or phone *inside* the bowl? Make a prediction and write it down. Then give it a try. What were your results? Were they the same as you expected? Were they different from the results with the sound source *outside* the bowl?

It's hard to see sound. But you can see its effects! Using a few common kitchen items, you can watch sound waves in action. Ready to hear some cool sounds?

Sound	Observation
Tap one fingernail on the side of the bowl	High-pitched, quick sound, no echo

QUESTION TIME!

How else can you "see" sound waves? What other surfaces might vibrate when exposed to sound?

Index

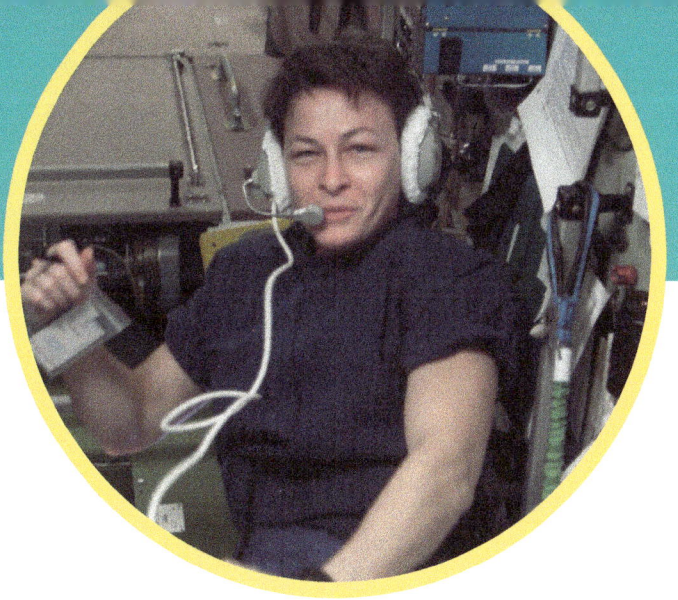

Glossary

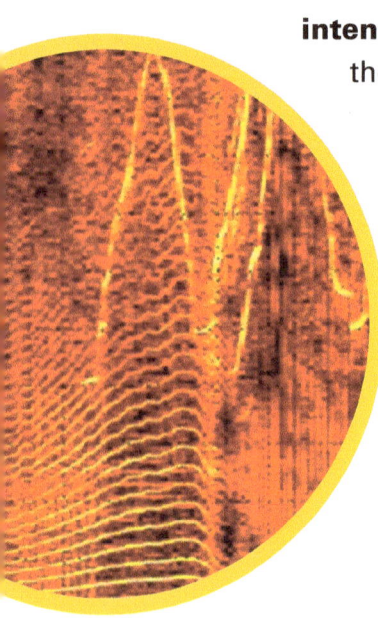

amplitude (AMP luh tood)—the height of a sound wave

compression (kum PREH shun)—when molecules are pushed closer together

diffraction (dih FRAK shun)—the bending and spreading out of light around an obstacle

echo (EH koe)—when sounds bounce back from a hard surface and repeat

echolocation (EH koe loe KAE shun)—gathering information on one's surroundings from reflected sound waves

frequency (FREE kwen see)—the number of waves that pass a certain point in a period of time

intensity (in TEN sih tee)—the amount of energy that moves through an area in the direction of a wave in a certain amount of time

medium (MEE dee um)—a substance that a sound wave travels through

membrane (MEM brane)—a very thin layer of tissue that covers organs or cells

period (PEER ee ud)—a given amount of time

pitch (PICH)—the highness or lowness of a sound

rarefaction (rare uh FAK shun)—the part of a sound wave in which the molecules bounce back after the initial compression

refraction (ree FRAK shun)—the action of sound waves bending and changing speed

sonic boom (SAH nik BOOM)—the loud noise made when a vehicle travels faster than the speed of sound, breaking the sound barrier

vacuum (VAH kyoom)—a space with no matter in it

www.ingramcontent.com/pod-product-compliance
Lightning Source LLC
Chambersburg PA
CBHW052142170526

45159CB00017B/3139